G000164083

1
GUIDE
TO BEING
A GOOD
HUMAN
FOR THE
EARTH

All proceeds help fund
Good Human projects.

www.goodhuman.eco

THE
GUIDE
TO BEING
A GOOD
HUMAN
FOR THE
EARTH

How to Live Sustainably

TRAVIS RAMSEY

San Francisco, California
www.goodhuman.eco

All proceeds go to fund Good Human projects.

Learn more at www.goodhuman.eco

First edition April 2020

ISBN: 978-0-578-63833-1 (paperback)
      978-1-7345552-0-2 (ebook)

Library of Congress Control Number: 2020901222

GOOD HUMAN
2443 FILLMORE STREET #380-6274
San Francisco, CA 94115
www.goodhuman.eco

Printed in the United States of America

Distributed by IngramSpark

10 9 8 7 6 5 4 3 2 1

TABLE OF CONTENTS

QUICK INTRODUCTION

THIS IS A GUIDE to being a good human for the earth. How can we live on this planet sustainably and have a good quality of life for many generations to come? This book provides quick and practical information on how to live sustainably. The first section, "Foundation of Living Sustainably," gives you an overall understanding of how you can think about living sustainably and why. The second section, "The Guide to Living Sustainably," details straightforward things you can do to live more sustainably. The third section, "Sustainable Concepts You Should Know," provides a brief explanation of terms, concepts, organizations, etc. related to living sustainably that you should be aware of. Lastly, you can take a quiz to calculate your Good Human Score to see how well you are living sustainably and where you can improve.

I believe that we can live in balance with the earth and maintain a healthy biodiversity without sacrificing our quality of life. The solutions and knowledge are already here—all we need is broader awareness and the will to change. I hope this book can bring more awareness to the general population and inspire change.

Go to www.goodhuman.eco website for updated information, as well as additional resources.

FOUNDATION OF LIVING SUSTAINABLY

THE FOUR PRINCIPLES OF SUSTAINABLE LIVING:

Consume Less, Do No Harm,

Buy Renewable, Cause Least Impact

To live in a sustainable way is to live your life so you are not causing future generations of the current living organisms (humans, plants, animals) to live a lower quality of life than you are today. There is a large amount of information and products out there that advertise themselves as "eco-friendly." This topic is vast and touches every aspect of your life, from what you eat, your housing, the clothing you wear, transportation, etc. There are many certifications out there and more coming out. I want to provide you with a framework of thinking that you can apply logically to all the decisions you need to make in order to try to live more sustainably.

Every day, we make choices that support the current system, which is ultimately unsustainable. As individuals, we have the power to make choices in our buying decisions that will influence businesses to support our demands. You vote with your dollars for the change you want to see. Most importantly, you should support and vote for government policies and political candidates that will help achieve sustainable living.

If you wish to live sustainably for literally everything you do or consume, you need to ask yourself these four questions, in this order:

Do I need this?

You should use and consume only what you need. Be strict about everything you buy and question if you really need it. Buy less stuff. Reuse what you have or buy used. Buy quality products that are long lasting rather than cheap things you plan to just buy more of later. Don't waste. Eat all your food. Repair your stuff instead of replacing. Buy with intention and consume less.

Will it harm me or the earth?

Choose products that do not contain any harmful materials or chemicals for human health, animal life, or plants. This is especially true for things that you eat. Food produced with fertilizers harm the environment, so buy organic food at a minimum. When possible, eat sustainably grown food (also known as regenerative farming), and buy natural materials made without toxins. Does this product's manufacturing process or its sourcing of materials harm the earth?

Is it renewable?

Buy all materials, products, and energy from renewable sources. For example, use solar and wind power for energy, since gas is in limited supply and warms the climate with CO_2 emissions. Buy wood from sustainably certified sources (including any products that contain wood, like a chair). Ask yourself: Is this product or material recyclable or compostable? Think about the whole life cycle of the product and all sourced materials used in the product.

Does it use the least resources?

Buy what is produced with the least waste, energy, and impact on the environment. For example, concrete uses more energy than wood, so wood products are preferable. Raising cows for meat and dairy requires significantly more resources than growing plants, so eat mostly fruits and vegetables. Choose products from local producers, since less transportation is needed and therefore, fewer resources are used to consume the product.

HOW TO ADOPT SUSTAINABLE LIVING

Adopting a fully sustainable way of life can be difficult, if not impossible. After all, we are all living creatures, and living creatures inherently make their impact in the world. It is the nature of life. The goal is to have the least impact or at least, the impact you desire.

When you begin the effort to live sustainably, start with things you can do that provide little to zero inconvenience to you. Some things just require knowledge and forethought and are easy to do. Other changes require you to spend more money or time. Obviously, you should start with changes that only cost a little more, and work your way up as best you can for what you can afford in money and time. When you struggle to accept something at a higher cost, remember that prices of the things we buy are not reflecting their true cost if they are not produced, distributed, consumed, and recycled in sustainable methods. With this knowledge in mind, you can choose to pay closer to the real price. Hopefully, one day, government regulations will force all prices to reflect their true cost, or we eventually will be forced to accept this cost due to exhaustion, depletion, or damage of our natural resources.

The next step is to make sacrifices in what you are used to. Some changes are of little inconvenience, such as using reusable shopping bags. Other things are big sacrifices that may be hard to accept that your life can no longer have.

You do not have to save the world tomorrow. The idea is to do what you can *now*, however little or big the effects are. Any change made by individuals helps. Some may want to do more but feel that they can't with the current conditions. It's true: that with our modern culture, economy, and manufacturing processes, it is nearly impossible to live completely sustainably. However, the more demand and awareness we create, the more we influence government and businesses to provide the solutions we need and want, which will create more options for choosing sustainability.

When considering options for living sustainably, we can score those options in the following categories:

Impact

"What is the impact for helping the world live more sustainably by adopting this sustainable option?" Some things make little impact and others make a huge impact. The biggest impact you can make is to support government policy and candidates that help achieve sustainable living. The next is doing anything that stops emitting CO_2 emissions, including burning fossil fuels and eating meat. Third is reducing energy consumption by doing less transportation and getting energy-efficient appliances and lights.

Money

"How much more do I need to pay compared to non-sustainable options?" If it doesn't cost that much more, we can more easily adopt it. Opting in for clean energy with your energy provider, for example, costs little and immediately stops supporting fossil fuels for your home energy. But buying an electric car might be more expensive. Eating all organic or sustainably grown food or sustainably produced clothing might also cost too much now, but hopefully that will change with demand.

Effort

"How much effort must I make?" Some things require little effort, like using reusable shopping bags, while others take more effort, such as composting your food waste when you have no city composting services. Not wasting your food also requires little effort but makes a major impact on conserving resources. Using anything reusable takes some effort and so does repairing your stuff but helps conserve resources.

Time

"How much extra time will this take to do?" Some things take little time, such as calling your energy provider to opt in for clean energy. But taking public transportation to work or biking might take more time than driving but would save CO_2 emissions.

Sacrifice

"How much of a sacrifice in my quality of life will this

be?" You might love eating meat, for example, but eating meat consumes a great deal of resources and influences climate change. So, eat meat as little as possible. Also, just buy less clothes and stuff in general.

When making sustainable choices, you are often prioritizing different principles of the sustainable mindset. Do your best to find an option that balances what you can pay or sacrifice with sustainability.

WHY SHOULD YOU CARE TO LIVE SUSTAINABLY?

Why live in a sustainable way? Good question! The most obvious explanation is to not ruin our environment, not destroy all the trees, and make animal life go extinct. Let's keep the environment as it is with the current systems and species. Let's preserve the environment and not make a negative impact on the planet. I think all these things are great, and any environmentalist would agree. This is the reason provided by most people that are currently promoting and pushing for change. However, unless you truly like to go hiking and see nature beyond human life or if you like to eat food that is harvested from the wild, then you might not care about saving it. To protect the environment requires you to make the actions of living sustainably out of the goodness of your heart for the most part.

A more selfish reason that could convince more people is this: if we change our planet enough, we may cause our own extinction. If, for example, we pump enough CO_2 into the atmosphere that it changes the climate too quickly, many species won't be able to adapt quickly enough, and we may put our own human existence into jeopardy. Humans have evolved into our current environment, and our ability to exist depends on the planet to be in a particular state for us to have access to vital resources, like water and food.

Of course, the world is always changing and evolving, and it will continue to do so. For example, the atmosphere didn't always have this level of oxygen, and the overall temperature in the Ice Age was much colder. But if it changes too fast, we may not be able to keep up. So, we *must* adopt a more sustainable way of living to save our own species. If we change the environment to a point that makes humans go extinct, the planet will still be here, and I'm sure some life will evolve and keep on living without us. History teaches us that this is common. There have been a few mass extinctions on this planet, and each time the planet changes and life evolves, but we have to be smart enough to at least not cause our own extinction. We may not be able to prevent natural disasters or a comet from hitting the earth, but we can do other things to prevent our demise.

The choices we make today might not affect your lifetime, but it *will* affect your children and their children. I imagine there are some people without children or who don't wish to have them that might not care about humanity's continued existence and I cannot necessarily judge anyone who thinks this way. My only comment is that humans are pretty cool creatures, and even though they can be selfish and do harm sometimes, many of us do beautiful things. If you have a negative outlook on the human race, try to think of our potential and what future humans could be like and not simply what they are now.

However, despite how well we take care of the planet, the unfortunate reality is that our sun—upon which all life depends—is going to burn out in the distant future. The sun is this planet's primary source of energy, and we need some form of energy to survive. Plants grow from the sun's energy and we eat the plants; the sun's energy evaporates water and makes clouds to create rain. Even wind is made from the sun due to temperature fluctuations in the rotating planet and warm air interacting with cool air. Without the sun, all life perishes. All stars, including our sun, have a lifespan and will eventually die (just like all other things), so there is a deadline of when this planet will not be livable.

So, what is the point, then, to maintain our planet in a condition for us to live, when in the end, all life on earth will end when the sun enters the last stages of its life?

The first reason is because life itself is so beautiful and awesome, so let's keep enjoying it for as long as possible and in the best conditions possible. Let's pass this gift of existence to our kids and their kids. The other reason is to buy us enough time to learn about the systems of the planet and life and to progress enough in science and technology to be able to leave this planet in a spaceship and move to another star. There are millions of stars out there in different phases of their life cycle, so the solar energy we get from the sun is available elsewhere. We can use this energy to create an ecosystem on a spaceship, and then travel to another star and maybe even a new planet.

Unfortunately, with modern science, we do not know how to do this yet, and maybe we never will, but we can try! We need time to study, advance science and technology, and figure it out. Every day, we gain collective knowledge quickly and make a great deal of progress.

Bringing this all full circle: one of the first and most important steps to building a spaceship that is self-sustaining is to understand how we can live on this planet sustainably. We need to learn how to live by recycling resources, to live in a circular and regenerative way. We must master this so we can recreate this system on a spaceship and move to another star. Even Mars, which some think could be our next home, is not a viable long-term option since its main source of energy is also our sun. No matter where we go, learning how to live in a sustainable way and harnessing and making the best use of extrasolar energy is the key to mankind's continued survival.

Maybe you think that humans only belong on planet Earth and should die with this planet. Yes, we were born out of this Earth, and of course, this is our ancestral home. We only exist the way we are because of the things that have happened in the past and the conditions leading to our existence. But all life forms seek to continue their existence wherever they can. They try to live in all conditions and test their boundaries. There may even be some life on our planet that has traveled from elsewhere. Seeds from plants drift in the wind and land all over in an attempt to simply exist. It is the nature of all living things to try to continue to exist and evolve. Therefore, our attempt to leave our dying sun to attempt to exist is natural. We don't know what the future holds, but we can hope for the best and make our best effort.

THE NEED FOR A STRONG WORLD GOVERNMENT

We need a strong world government to define and enforce basic laws over the whole world. These laws should protect basic human rights and environmental laws, ones that affect the world or surrounding regions, not necessarily only local damage. Examples include CO_2 emissions and plastics into oceans, as well as basic human rights like safety, shelter, food, and water. We need to adopt a mentality of global citizens who stand for these basic needs of humanity and respect all other aspects of local cultures and encourage and support our unique expressions in different regions. This global government should not be optional or by invitation.

The reality is that the pollution of burning oil in one country goes into the atmosphere that the whole world shares. So, one country could pollute "their" air and affect the entire planet. Or a country can dump plastic into the ocean, and the ocean currents carry that plastic all over the globe. Unfortunately, we do need to enforce global environmental regulations in order to protect all people in all countries.

Another issue is making a country stop polluting the earth by burning oil when they are poor and struggling, while other countries have increased their wealth by burning oil the last couple hundred years. To make it fair, wealthier countries need to provide support. Countries that have damaged the earth need to pay the actual cost of the damage.

THE MOST IMPACTFUL THINGS YOU CAN DO

The top three easy and most impactful actions you can do now to help the world live more sustainably:

1. Vote for sustainable living policies and candidates.

2. Call your energy provider and opt in for renewable energy.
3. Eat less meat.

Studies show that we are on track to warm our climate enough by 2030 to make major changes to our ecosystem. Sea levels will rise, which will require coastal areas to adjust or move, causing suffering and expense. Our current ecosystem will change too fast for species to adjust and will cause many extinctions. We depend on a diverse ecology in order to have a healthy planet to survive. So, the number-one priority is to stop warming the climate. Here are the most impactful things you can do to help.

Support Government Policies

Advocate, support, vote, protest, and promote all government policies that will support sustainable living goals. Do the same for political candidates who support such policies and will help implement them. Unfortunately, individual actions alone to make your own life sustainable are not enough and there are many people who are not aware or do not care about living sustainably.

The only way we can truly make enough change is through government regulation. We need regulation to ban the use of fossil fuels, single-use plastic, unsustainable farming practices, and more. The government needs to force the changes all humans need to make in order to protect our own existence. The government and our tax dollars need to be spent on restoration, regeneration, and sequestration projects in order to bring carbon out of the atmosphere and restore our environment to a healthy state. Governments must focus on converting all energy to clean energy and sequestering carbon. One effective way to sequester carbon is utilizing regenerative farming practices. Longer term projects include moving to more clean energy public transportation, as well as walkable and bikeable communities.

Use Clean Energy

Call your energy provider today, and ask if you can opt in for clean energy. This is easy and doesn't cost much more than you already pay. By doing this, you pay for your energy that comes from clean renewable energy, such as solar and wind. You also build demand and resources for the utility provider to move all their energy generation to clean energy. If they don't have this option or in addition to this, have solar panels installed on your property. When your appliances are ready to replace, buy all electric appliances, such as the stove and water heater. For your next car, buy an electric car or plug-in hybrid. *We need to stop burning CO2.*

Eat Less Meat

Unfortunately, a cow takes many resources to grow, including land and water. They also release methane, which warms the atmosphere more than CO2 emissions. I'm not saying we all have to go vegan, but the less meat you can eat, the better. Buy meat from sustainable and local sources. The cheap meat industry is a huge issue in the world. Forests are cut and burned down to make space for grazing cattle or to grow cheap feed for cattle. Do not support the cheap meat industry, such as fast food. Enjoy a nice steak every once in a while if you like it. Eat a small amount and not too often. If you can, don't eat meat at all.

Buy Food Grown with Regenerative Practices

The agricultural industry has one of the biggest impacts on our environment. It uses land, water, and energy. "Conventional" farming practices use fertilizers that are mined and in limited supply to provide nutrients to single-crop fields. Fertilizers also contaminate our environment and ruin the soil and water. One day, we will run out of the mined ingredients used in fertilizers. So, it's better to transition now to regenerative farming practices.

Regenerative farming uses compost as a natural fertilizer, which is basically broken down organic matter, like food scraps. It also uses cover crops to provide a natural cycle and replenishing of nutrients into the soil.

The biggest thing is to stop tilling the soil, which means to churn it up each year. This is huge because when you allow the soil to be rich, with long root systems and organic matter, it sequesters carbon from the atmosphere. Given the amount of farming we do, certain organizations like Patagonia think that we could sequester enough carbon out of the atmosphere to prevent climate change by adopting regenerative farming. That means buying organic food at a minimum, and then look for any food using sustainable, regenerative, no-till farming practices.

Reduce Waste and Consumption

The last most important thing is to generally reduce waste and consumption of everything in your life. Take this idea and apply it to everything. Use energy-efficient appliances and LED lights. Turn your lights off when not needed. Use less air conditioning and heating. Weatherize and seal your home to use less energy to heat and cool it. Don't waste food. Recycle and buy recycled products. Compost your food waste. Use reusable everything as opposed to single-use items, such as reusable shopping bags, water bottles, and food containers instead of ziplocks or beeswax instead of plastic wrap.

If you must use paper plates for a BBQ, use compostable plates. Buy any products that have reusable containers and packaging. Drive less and walk or bike or take public transportation. Buy things that are produced locally to reduce transportation. Repair your shoes and clothes to use as long as possible. Buy fewer clothes. Minimize water usage by replacing your lawn and planting native plants. Install low-flow shower heads and faucets. Buy water-efficient dishwashers and toilets. These products are readily available today and they work.

GUIDE TO LIVING SUSTAINABLY

OPT IN FOR CLEAN ENERGY WITH YOUR ENERGY PROVIDER

Call your energy provider, and ask to get your energy from renewable sources. Not every energy provider will have this option, but you should call and ask. If they don't, urge them to provide it. Certain energy providers will provide you with different levels to opt in for clean energy. Go with 100% renewable energy; the extra cost on average is only $3.29 per month!

Opting in for clean renewable energy from your energy provider allows you to use solar and wind electricity at your home or office without having to install solar panels. You will be supporting projects to get more renewable energy and help with the transition, and it goes a long way in limiting CO_2 emissions while providing electricity needs for your home.

GET SOLAR PANELS OR WIND TURBINES AT YOUR HOUSE

If you are already hooked up to the electricity grid, the best and easiest thing to do is call your energy company and opt in for 100% clean renewable energy. This will help fund the electric company to build more infrastructure for renewable energy. It is a more efficient use of resources to build larger scale solar and wind farms than small individual ones at your house. However, if your electric company does not have a renewable option or if you want better control and backup of your own energy usage at home, then install solar panels or a wind turbine to generate your own clean electricity.

Another benefit of installing solar on your roof is that you can utilize roof space for something beneficial. The energy company needs to lease large tracts of land to set up solar panels. But solar panels on every building requires no extra land.

Many cities and states offer rebates and tax incentive programs for installing solar panels. Contact your local solar or wind turbine installer, and they can provide the full breakdown of costs and long-term payoff. Often, you can even get financing for the upfront costs, and it just ends up saving you money in the long run.

When you install solar, you can also opt to have it hooked up to the electric grid to sell your excess energy back to the energy company. This is great because you make money on your extra electricity, but you also use electricity from the grid if needed. However, in some scenarios, this still makes your electric system reliant on the grid since it is processed through their system. This means if the grid goes down for whatever reason, then you can't just use your own electricity without changing it. Be sure to ask these questions and understand the pros and cons before you decide.

REDUCE ELECTRICITY USAGE

Think about anything you can do to reduce your electricity usage. It can be as simple as turning off lights when not in use. Always think about appliances and devices being on and if they really need to be. Buy energy-efficient appliances. Configure devices to turn off when not in use, such as your computer or TV. Turn off your printer when not in use. Configure your AC and heater to a wider range of temperatures to use less energy. Set up home automation to automatically turn off lights when not in a room or turn off AC/Heater when not in a room. Have a master switch at your front door that will turn off all outlets in the house. Just having outlets on and appliances plugged in takes energy. Every amp of energy you use consumes resources.

WEATHERIZE AND SEAL YOUR HOUSE

Heating your house or other building is one of the biggest energy consumers; however, with certain improvements, you can use less energy to heat your house.

Add more insulation to need less energy. You will save in the long term by needing a much smaller heater appliance and spending less energy to heat your house. Insulation is in the floor, walls, and ceiling. A green roof with plants growing on it provides insulation and looks good. Cellulose is another good option for insulation, since it is a natural material (wood/paper), and you blow it in, which fills all the gaps and spaces better than the batts (pads).

Seal your house well so there are no leaks of heat. You can wrap your house in a material that prevents air leaks, but if you want to improve your existing building, you can test for air tightness and then fix any leaks found. Use caulk and weather stripping to fill in any gaps or holes.

Replace your windows with double pane windows because these will not lose as much heat.

If you are building a new home or building, you can implement these strategies in the design phase, but you can also remodel an existing building to make these improvements. Simply add weather strips around doors and caulk areas where there are holes.

You can hire an inspector to do an evaluation of your home and provide recommendations. HERS is one standard in rating your home's energy efficiency.

USE LESS AIR CONDITIONING AND HEATING

Air conditioners, fans, and heaters are typically the biggest energy consumers and the biggest opportunities to decrease our energy usage. Make sure that you are not running the AC or heater when you are not home. Heat or cool only the room you are in rather than the whole house or building. Get a smart thermostat that helps you auto set temperature based on occupancy. Set your temperature levels at the minimum levels you can stand. Every degree helps save energy. If the room is cooler, put on a sweatshirt instead of turning up the heater. Open your windows to get your room cooler rather than use the AC.

USE LOW-ENERGY APPLIANCES

1. Induction Stove
2. Oven
3. Refrigerator
4. Dishwasher
5. Washer/Dryer
6. TV
7. A/C
8. Heater

9. Water Heater
10. Computer Monitor
11. Printer

When it comes to household or business appliances, go for energy-efficient options. Look for Energy Star-certified appliances, or check the actual watt usage required to run it. Look for features that will automatically turn it off when not in use.

For the stove, get away from gas and go electric. The best option is an induction stove because it uses the least energy and you can control the temperature better than an electric stove. The common complaint between electric and gas is that you can't turn down the heat quickly on an electric stove. Induction stoves take care of this issue. It does, however, require the use of magnetic-based cookware like cast iron or magnetic steel.

When it comes to TVs, the best option is an OLED TV. LED TVs actually use less energy than OLED, but OLED provides a much better picture quality.

Keep in mind that if you have a functioning appliance, it is better to use it until its life is over because the embodied energy it took to create that appliance is greater than what you will save by replacing it. It's best to replace the appliance when it actually needs replacing. If you can sell or give away your appliance to someone else to use so you can get an energy-efficient one, that works too.

USE LED LIGHTS

When it comes to lighting your home or business, LED lights are the only way to go. They use 75 percent less energy and last 25 times longer. They used to be more expensive but are very affordable today. Besides using less energy to manufacture lights since you don't need to replace them as often, you also save on the energy bill. As a bonus, LEDs can also deliver any mood and color of lightning.

LIVE IN A LEED-CERTIFIED HOME

When looking for a home to rent or buy, look for a LEED-certified home or a similar type of certification, like Living Home Challenge. These third-party certifications verify that the home is made to be energy and water efficient and is built with healthy materials and, depending on the level, more sustainable materials. Basically, they are built and/or operated more sustainably. The certification requires the same elements discussed here. If a home is not certified, you can also make your own assessment if it at least has a few elements of sustainability. There are different certifications and different levels. Obviously, the higher level is, the better.

EAT LESS MEAT

Unfortunately, a cow takes many resources to grow, including land and water. According to Water Footprint Network and other scientific studies, over 2,500 gallons of water are used to produce one pound of beef, compared to 477 gallons to produce one pound of eggs. Approximately two to five acres of land are used for each cow. They also release methane, which warms the atmosphere more than CO_2 emissions.

Buy meat from sustainable and local sources. Buy pasture-raised meat because better sustainable practices are used to grow them, and it is more humane to the cows. The cheap meat industry is a huge issue in the world. Forests are cut and burned down to make space for grazing cattle or to grow cheap feed for cattle. Do not support the cheap meat industry, such as fast food. Buying better, more sustainable meat certainly costs more because the unsustainable feed-lot type meat is damaging our environment. If you eat less meat overall, you will save money for the extra cost to eat meat. I'm not saying to go vegan, just that the less meat you can eat, the better. Enjoy a nice steak every once in a while if you like it. Eat a small amount and not too often. If you can, don't eat meat at all.

Keep in mind that animals are important for regenerative farming practices. Animals can be used to eat weeds and cover crops, and their poop is amazing natural fertilizer. Our natural ecosystem has animals and plants, and they work together to support a circular system. So, we should not remove animals from our food system—in fact, we need them to grow food in a sustainable way.

DON'T WASTE FOOD

Food waste is one of the biggest issues contributing to our impact on the planet. There is so much energy, water, labor, transportation, and packaging that goes into producing and delivering food. When it is wasted, we also waste all that energy. If you can waste less food, then you will use fewer resources and have less impact on the planet.

Try to buy only the food you need that will not spoil before you can eat it. Make the effort to eat all the food you make before it spoils. Save your leftovers from restaurants or when cooking from home.

EAT SUSTAINABLY GROWN, REGENERATIVE FOOD

The agricultural industry has one of the biggest impacts on our environment, as it uses land, water, and energy. "Conventional" farming practices use fertilizers that are mined and in limited supply to provide nutrients to single-crop fields. Fertilizers also contaminate our environment and ruin the soil and the water so we have to rely on fertilizers. One day, we will run out of the mined ingredients we use in fertilizers. So, it's better to transition now to regenerative farming practices.

Regenerative farming uses compost as a natural fertilizer, which is basically broken-down organic matter like food scraps and animal poop. It also uses cover crops to provide a natural cycle and replenish nutrients in the soil. The most important thing is to not till the soil, which means to churn it up each year. When you allow the soil to be rich with long root systems and organic matter, it sequesters carbon from the atmosphere. It also provides a healthy micro ecosystem in the soil, which is what healthy plants eat from. With how much farming we do, certain organizations like Patagonia think that we could sequester enough carbon out of the atmosphere to prevent climate change by adopting regenerative farming.

Planting diversity instead of single crops or mono culture provides resilience to the farm, protecting it from any invasive species and diseases because the variety is a natural protection. Diversity provides a cycle of nutrients because different types of plants use different nutrients from the soil and return other nutrients, creating a balanced and circular system.

Animals are an important element in regenerative farming because they can eat weeds and cover crops and their poop is an excellent natural fertilizer. Any sustainable farm needs to have animals incorporated to keep it healthy—there doesn't need to be an overabundance of animals, just enough to do their job in the circulatory system.

Eating seasonal foods is also more sustainable. You can get food that is more local because you are buying in season, rather than buying strawberries in winter that are flown in from Mexico, for example, which uses a great deal of energy in that transportation.

Organic food is better than conventional since it doesn't allow the use of many chemicals, but it certainly is not enough. So, buy organic food at a minimum, and try to buy any food that is using sustainable, regenerative, no-till farming practices.

You can look for grocery stores that have these in their mission for sourcing food, or buy direct at a farmers market and ask them about their farming practices. You want to hear any of the following: organic methods, compost, cover crops, no till, regenerative practices, sustainable practices, no fertilizers, no chemicals or chemical pesticides, diverse plants instead of monoculture, and efficient water management.

BUY SUSTAINABLE FISH

When you buy fish, make sure to buy from a sustainable source. Look for fish recommended by the Monterey Bay Seafood Watch organization. They provide lists of recommended seafood. Farmed fish can be a good source as long as the farming practices are sustainable. Wild fish are also fine, as long as the population of fish is not impacted due to the fishing or other parts of the ecosystem being impacted. For example, large-scale net fishing catches other sea life and kills them. The species is important, and so is where it comes from. For example, salmon is plentiful in Alaska, but not in California. You can also look at grocery store ratings like Whole Foods or buy from a store that holds these values.

This is important because if we don't buy sustainable fish, we will consume their populations and no longer have those fish not only for our own food but also for a balanced ecosystem. Right now, bluefin tuna is on its way to becoming completely fished out. Consider that next time when you go for sushi. But even if you stop eating bluefin tuna, that doesn't stop the person next to you from eating it, so we need government and world regulation to protect our food and wildlife systems.

BUY LOCAL

You should buy food and any other products that are more local because it uses less transportation energy to get to you. Buying local food and products also supports the local economy and makes your region less reliant on the global food system. However, it is more important to buy sustainably made products and regeneratively grown food from a good source rather than local, if you are deciding between the two. For local food, it is a good idea to shop at farmers markets.

BUY IN SUSTAINABLE PACKAGING

When you buy food and products, consider the packaging. The best is to get packaging that is reusable. For example, you can buy milk from Straus in a glass bottle and then return the bottle to the store so they can wash and reuse the bottle. This is better than recycling or composting packaging. Also, find a way to use reusable packaging or buy in bulk and refill products. For example, at some stores, you can refill cleaning products with your own containers or buy seeds from the bulk section—any way to prevent more single-use packaging. Buying in bulk is also better in general since it is less packaging.

The next best choice is compostable packaging, since those materials are put back into the food system for farming. However, you must have access to commercial-grade composting facilities to compost packaging. Paper and cardboard are also good options since you can recycle them and they are light so take less transportation energy. Other alternatives include aluminum, since it is also easy to recycle and light; glass, since it is easy to recycle, though heavy; and hard plastic, which is lightweight and can be recycled, though not as much as the other two.

EAT AT RESTAURANTS THAT SOURCE SUSTAINABLE FOOD

In our efforts to live sustainably at home, we want to support and demand those same practices in all the businesses we give money to. When you go out to eat at a restaurant, it is harder for you to control their practices and where their food came from. So, try to go to restaurants that source their food from sustainable and local sources. They will likely mention this approach on their website or in their menu. Look for organic food at a minimum. For meat, it should be grass fed. Seafood from sustainable sources should be approved by Monterey Bay Seafood Watch. Typically, if a restaurant mentions what farms they get their food from, it is a good sign that they care about their source of food and are trying to support a local and good farm. Otherwise, they try to buy the cheapest food ingredients they can, and those sources could be questionable.

BUY LESS STUFF

Everything you buy requires energy and resources to produce and deliver to you. One of the quickest ways to make an impact is to reduce your consumption. The most sustainable product is the one you don't buy. With every purchase you make, consider if you really need it. Buy things that you really need, and buy good quality that will last a long time. Repair your stuff to make it last longer.

We need to shift our consumer culture view to have different values. With the discovery of oil and innovations, we have been able to produce many products cheaply. We have gotten used to being able to buy a lot of products and developed our consumerist culture. Another issue is our capitalistic economy. The incentive of people and companies is to make profit. You need to sell stuff to make a profit. We report daily on the health of our economy with Gross Domestic Product (GDP), which measures how much stuff we produce. We need to value living with less.

Try to avoid looking at advertising. The advertising industry is good at making you feel that you want something. If you are not exposed to this advertising, you will not have those feelings to buy things you don't necessarily need.

REPAIR YOUR STUFF

Try to buy things that are repairable and high quality instead of things that might be cheap and considered disposable when they break. Take your stuff that breaks or is damaged and get it repaired. Some things you can repair yourself, such as a hole in your shirt that can be sewed up. Or, order a new part and replace it yourself or take it to a repair shop and pay a professional to repair it. Shoes, for example, you can easily replace the bottom sole, inside cushion, and laces and keep using them. Repair your furniture, appliances, electronics, etc. Even if it is not cheap to repair something, it should at least be cheaper than buying a new one, and you have conserved the resources it takes to produce a new product.

Replacing a part on a product takes less of the earth's resources than building the whole product.

BUY USED PRODUCTS

Whenever you need to buy something, especially an item that is usually more expensive and lasts a long time, try to find it for sale used. When you buy a used product, you are extending its life. When you buy new, you are responsible for all the new energy and resources it took to buy the new product. A used product adds no more impact on the environment, and of course, you can save money as well. So, the next time you are buying a car, roof rack, snowboard, desk, coffee table, TV, exercise bike, etc., check the classifieds on Craigslist. org, Facebook Marketplace, and Next Door first. Or, buy used when you see an option on Amazon.

SELL OR GIFT YOUR STUFF

If you no longer want a product that you own, sell it or give it away instead of throwing it in the trash. This way, the item is still used instead of someone needing to buy a brand-new one. Every product requires energy and resources to produce, so the fewer things we need to produce new, the better. You can list your stuff for sale at market price or just at a low price to get someone to take it off your hands. Either way, it's much better than putting it in the trash. You can also give it away either on the Craigslist Free Stuff category or take to a Goodwill. Taking it to Goodwill allows you to write off the donation as a tax deduction.

BUY SUSTAINABLY MADE CLOTHING

The most sustainable clothing is the clothing you don't buy. Before buying anything, consider if you really need to buy it. Buy your clothing with intention. You'll want to buy from companies that support sustainable practices, which means they source their materials from sustainable sources and produce the clothing with sustainable practices and humanely. Good materials for clothing are hemp, bamboo, cotton, linen, and wool. But the source must be from a regenerative farm or at least be organic.

Rayon and modal are made with natural materials like wood pulp but require harmful chemicals to convert it to fabric. Polyester is made from plastic, which is made from oil. Also, when you wash clothing made of polyester, micro plastics wash into the water system and into the oceans and potentially, our drinking water!

You also need to know how the clothing is dyed. The blue jean industry dumps toxic dyes into the waterways in Asia. You want your clothing to have natural dyes.

Buy high-quality, durable clothing so it will last longer and you will buy less of it. Spend the extra money for sustainable clothing and it will also last longer.

Buy from producers that are more local so you support the local economy and less transportation energy is used.

BUY PRODUCTS MADE OF SUSTAINABLE MATERIALS

1. FSC-Certified Wood
2. Bamboo
3. Hemp
4. Cork
5. Eucalyptus
6. Wool
7. Organic Cotton
8. Leather
9. Glass
10. Clay
11. Metal

When you buy anything, check what is it made of and pick products that are made with sustainable materials, which means they come from a source that can be renewed, reused, and recycled. Anything made of organic matter is good, since it can be renewed by growing more. But it has to come from a source that can be renewed as it is consumed, rather than for example, chopping down trees in a forest that took 100 years to grow and consuming that forest faster than it can renew itself.

If you are buying things made of wood, look for

items certified by the FSC, which is the Forest Stewardship Council non-profit organization that certifies wood sources as sustainable. Wood is a great material because it stores carbon out of the atmosphere, can be recycled back into the environment, and is non-toxic. Bamboo and eucalyptus can be a good sustainable wood sources because they grow easily and fast. Cork is also a good option. Hemp is another great material because it grows fast and is easily renewed.

Other natural materials are wool, leather, and cotton. Leather is not good to use in high quantities and should just be taken from the animal food industry, not from animals grown just for leather. But it's good because it's natural and can cycle back into the environment.

If you are not buying things made of organic materials that can decompose back into the environment, then you want to select materials that have been reclaimed/reused or recycled and have the ability to be reused or recycled again. Steel is generally fine because it can be reused and recycled many times. Glass is also fairly easy to recycle and can go back into the environment naturally without harming it.

USE REUSABLE SHOPPING BAGS

This one has been around for a while and getting more and more normal. Some cities have banned plastic bags altogether, which is a great idea. Some charge you 10 cents to get a plastic bag to try to incentivize you to bring your own. So, you should buy reusable shopping bags and remember to take them to the store. This saves on the resources and energy required to produce all those plastic and paper bags. Just because you can recycle paper bags and even sometimes plastic bags does not make it okay. It still takes energy and resources to recycle. A reusable bag is produced once, then used over and over again for a long time and consumes fewer resources. This one is a no brainer. But it's important that you remember to take them with you when shopping. Try to create a process so you don't forget them. If you shop with your car, for example, unload your groceries and then put the bags back in your car right away or by your front door so you remember to take them. It's a bigger issue if you continue to buy more reusable shopping bags to a point that you have more than you need, since it takes more resources to produce a reusable shopping bag than a disposable one.

USE REUSABLE CONTAINERS

Instead of buying plastic water bottles when you travel, buy a water bottle and refill it. When you get coffee, get it in a "for here" cup that the coffee shop can wash and reuse, or bring your coffee cup if you want to take it to go. Apply this concept to anything you buy that could use a reusable container instead of a disposable one. This will help save on resources to produce those single-use containers.

Here are more examples: Use tin tupperware containers to take your lunch to go instead of aluminum wrap or paper bags. Refill your soap dispensers instead of buying new ones. When you picnic or BBQ, use reusable plates and silverware.

USE RECHARGEABLE BATTERIES

Instead of buying single-use batteries that you need to recycle in a special way, buy rechargeable batteries and a charger. You probably use batteries for your wireless keyboard, mouse, remote controls, and maybe for camping gear like flashlights or fans.

A battery charger will cost you $10-$20, and AA rechargeable batteries cost about $2 each. You want to buy enough batteries for all the devices you need and then double it , so you always have a set of batteries charged and ready to replace. Once you invest in rechargeable batteries, you won't need to buy batteries again, so you will save money in the long run. Also, if you opted in for renewable energy from your power company, then you can charge those batteries using clean, renewable energy. You also save on the resources and energy to produce and recycle more batteries.

USE REUSABLE NAPKINS

We typically use a lot of paper napkins and paper towels around the house for meals and cleaning. Instead of using one-time-use napkins for a meal, buy cloth napkins and kitchen towels to reuse. For napkins, you should buy two weeks' worth of napkins—I suggest one napkin per day minimum per person to use as their napkin for the day. Then you can wash all the dirty napkins every two weeks. Any way to cut down on single-use items and reuse is the goal, and this is an easy thing to do.

PICNIC WITH REUSABLE OR COMPOSTABLE PLATES AND FORKS

If you are going to have a BBQ or a picnic, the best thing is to use reusable plates, forks, and spoons. You can buy lightweight plates at a camping store, like REI, that are made of metal and cost $6/plate. But they will last forever. It feels nicer to eat on a real plate as well. Take them on the picnic/BBQ, and then take them home and wash them in your dishwasher. Do the same for forks, spoons, and knives. With this strategy, you use less resources to produce all the disposable plates.

If you need the convenience of disposable plates and cutlery, then buy compostable plates and forks. Never buy styrofoam, as it can't be recycled and doesn't decompose. The plates/cutlery package will say if it is compostable. Typically, these are made out of paper and the forks are made out of bio-based plastics or wood. Compostable is better than recyclable because the food soils the plates and forks and then you can't recycle it anyways. You will need access to a commercial composter to compost bio-plastic forks. If your city trash company picks up composting, they should be able to compost these items.

USE PAPER-ONLY ENVELOPES

If you use envelopes, buy recycled paper-only envelopes without any plastic window. Some recycling facilities can still recycle an envelope with a plastic window, but you would have to check with your recycle company. But it certainly doesn't help since they need to have the technology to remove it. Some people take the time to cut out the plastic window, which is annoying to have to do. So, let's just not have any plastic windows on our envelopes. This way, they can definitely be recycled.

BUY AND RECEIVE SUSTAINABLE GIFTS

Take your sustainable living practices to gifts as well. Living sustainably, as you have learned, is about buying the right products and only buying what you need. Holidays and birthdays make us all buy gifts and support this consumer culture. Unfortunately, we give and receive many gifts that are not needed or wanted, and that waste makes an impact on the environment.

Here are a few ideas for sustainable gifts you can get for people or advise your friends and family to get for you as a rule:

1. Experience Gifts: Buy a ticket or gift card to a concert, show, trip, restaurant, or lesson/class.
2. Consumable Gift: Buy any food or beverage that will be consumed, like chocolate or wine.
3. Charity Gifts: Donate to a charity that you support.
4. Tree Planting: Donate to an organization that plants trees.

If you must buy a product, then follow the same principles of buying other sustainable products and things you know the person really wants and needs. Encourage them to return it if they don't need it or want it.

USE ECO-FRIENDLY CLEANING PRODUCTS

For all cleaning products, you want to get non-toxic products that don't harm the environment. This basically comes down to using more natural ingredients and fewer harmful chemicals. When you use soap and shampoo, these products end up going through our sewage system, water systems, and into the oceans. If you use harmful chemicals in your soaps, then you are polluting the environment. The same goes for laundry detergent and any cleaning products for the home or business. Look for products you can trust that advertise themselves as being good for the environment.

Buy cleaning products in bulk size with refillable bottles to use less packaging over time. You can buy one hand soap dispenser, then get large soap containers to refill it instead of buying a new plastic soap dispenser every time. The same goes for dish soap. If you can find a source in your neighborhood, refill your cleaning supplies in existing bottles. Some grocery stores have bulk cleaning products areas, similar to how they sell seeds and nuts in the bulk section, where you can fill up your reusable containers.

For bath soap, the most sustainable option is locally made bar soap made of natural organic ingredients. This takes less transportation to get to you and has no packaging.

Another benefit of buying cleaning products that are good for the environment is that it is good for your own health. There are seriously toxic cleaning products like bleach that contain harmful VOCs (Volatile Organic Compounds)—when you breath them in, it can cause health issues.

ADD A MICROPLASTIC FILTER TO YOUR WASHING MACHINE

If you have any clothing made of polyester or other plastic-based materials and wash it in a washing machine, little microplastics can shed off the clothing, go into the water system when the machine drains, and end up in our rivers and oceans. This is a big contributor to the pollution of microplastics all over the world. We are finding micro plastics even in the bodies of our sea life and drinking water.

Try to buy clothing that is made out of natural materials instead of plastic like polyester and nylon. Then you won't need to worry that you are polluting the environment with microplastics. If you must have polyester clothing, then buy from a good brand of good quality that hopefully will not shed microplastics as much.

You can also install a microplastic filter on your washing machine that will catch and filter all the microplastics out of the water before it goes into the drain. A device like this can cost about $140. There are also laundry balls you can buy that catch microplastics, which you just throw in with your clothes. The Cora Ball costs $40. It's best if you use both solutions to catch more fibers. Another option is getting a Guppy Bag for $30 that you put your synthetic clothes into and then wash with the rest of the laundry. This bag helps prevent plastics from shedding off your clothing and captures any that do.

COMPOST

In the natural environment, plants die and decompose back into the soil, which creates a nutrient soil that feeds microorganisms and new plants that are growing. When we farm our food, the plants go to our homes, and we eat them instead of those plants recycling back into the environment. What we can do is send our food scraps back to the farm to use for healthy soil. That is what composting is. You collect your food scraps and let them decompose, and then farms use this rich organic material on their soil to grow more food. This means they don't need to use chemical fertilizer to provide nutrients to their plants. Compost is also organic. There is so much energy in all the food waste and scraps we throw away into landfills that can go back into our farming system to make our farming more regenerative. We can't rely on fertilizers because those sources come from limited mined materials and will run out. We must farm in a regenerative way. Composting is a major component of that.

Hopefully, your city trash collector collects compost. Typically, it is a green bin. If you don't have one, check your trash collector website or call your city officials to ask about this. Every city must adopt composting as part of their trash collection to get this major organic resource back to the local farms. If your city does not have this, you can advocate and vote for it.

If you have a garden at home, you can compost at home and use it in your own garden. There are composting bins you can buy to help you compost, or you can create a pile that you rotate around until it is ready to use. Check out a book from the library to learn about composting or take a class. Another way to compost is to get a worm garden—you feed the worms your scraps to eat, and their poop is nutrient-rich material for your soil.

RECYCLE PROPERLY

If you can't compost something or reuse it, hopefully you can recycle it instead of putting it in a landfill. When you recycle, those materials get processed and reused. Remember that recycling is not the best solution overall because it still takes energy and resources to collect materials then process them again into new products. So, don't think that just because you recycle your single-use water bottles that everything is fine. Instead, you should be using a reusable water bottle and not buying any plastic bottles.

There is a pretty big issue with our culture recycling properly. Check with your trash company website about what you can recycle and become educated on how to do it properly. When you put things in the recycle bin that are not recyclable or are contaminated, then the whole batch of that recycled trash often gets rejected and goes to the landfill. So, all your efforts to recycle might go to waste if you don't do it properly.

Generally, you should be recycling clean and dry paper, cardboard, hard plastic, aluminum, metal, and glass. Then, depending on your recycling company, they will advise what and how to recycle other things, like aluminum foil and plastic bags. You can't recycle products that have mixed materials, like waxed cardboard or metal with paper to create a pouch, like toothpaste. One of the biggest issues is when things are soiled with food or water. Empty the food into your compost and wipe with a used paper towel before you recycle. Don't spend too much water trying to rinse everything clean. Soiled paper towels and paper plates should go in the compost. If you are not sure if you can recycle or compost something, it is better to put it in the landfill trash because you might contaminate the recycle bin. Hopefully, your trash company sorts through the landfill trash to pick out items they know can be recycled.

If for some reason your city still doesn't recycle, you must call your city office to demand that this service be offered. Otherwise, you'll have to find places to take your recycling to yourself.

PROCESS AND USE HUMAN WASTE

In the natural environment, everything is cycled. Animals eat plants and then poop. The poop decomposes back into the soil and provides nutrients to plants to grow again. Animals and plants die and decompose into the soil to provide nutrients to more plants and thus animals. The goal is to support this model of cycling of everything to create a circular system. One area we are losing many resources is in human poop and urine. We eat food, which takes resources, and then we go to the bathroom, which goes to sewage and is stored or put back into the environment after being treated. We can utilize the organic matter in human waste for other uses.

We can use human waste to generate electricity with biogas or hydrogen fuel. We can compost feces to use in farming or use urine as fertilizer. We can use feces as a food source for black soldier flies, which we can then use for feed for other farm animals and fish. We can make briquettes out of processed poop to use as fire fuel instead of wood.

If you are hooked up to the city sewage system, then ask your city or sewage processing company what they are doing with the human waste and advocate for utilizing it if they are not. If you have a septic tank, check with the company who collects the sewage. If you are not hooked up to central sewage processing, then you can look into solutions to install it on your property. Composting toilets are an easy one.

There are many inventions and people working on solutions for this, but it is not very mainstream yet. There are also serious precautions that need to be considered in order to do this properly to avoid any health issues. But we will need to figure this out and support these solutions so we can utilize this resource and support a regenerative circular system.

INSTALL WATER-EFFICIENT APPLIANCES AND FIXTURES

You can save a great deal of water by using water-efficient appliances. When you need to replace your toilet, look for one that uses less water to flush. Check the gallons per flush. Install low-flow faucets and shower heads. Low flow basically sprays less water out but can still maintain a good pressure. Installing an aerator on your existing faucets is cheap and easy. Seek out products with WaterSense certification.

When you buy a dishwasher, check how many gallons it uses for a wash cycle and look for one that uses the least. Seek out Energy Star-certified machines. Another tip to save water when washing dishes is to not rinse the dishes before loading the dishwasher. Just scrape food scraps into the compost and load directly into the dishwasher.

Another side effect of saving water is spending less on your water bill.

MAKE YOUR GARDEN WATER EFFICIENT

If you have a garden or yard, change it to have plants that use no extra water or less water. A grass lawn takes a great deal of water to keep the grass healthy. You can design your garden with native plants and pathways going through plants and trees instead of a lawn. If you want something like a lawn, consider fake grass. Go to your garden supply store or hire a landscape designer, and tell them you want to make your yard with water-efficient plants. Native plants of the area typically don't need extra water because they evolved to live in that region naturally with the usual rain. So, those are always good options.

Design your watering system to use the least amount of water. A watering system on a timer that is hooked up to the internet to know the local weather will adjust watering times if it is a rainy day. Use moisture readers to understand if plants need to be watered. Set up drip irrigation systems that drip water specifically where it needs to go—by the roots—instead of sprinklers, which spray everywhere on the leaves and evaporate in the air. Also, water in the morning before it gets hot.

REUSE GREYWATER

Greywater is the water that has been lightly used, like the water in your faucets, showers, and dishwasher. It is not toilet water. There could be traces of food or cleaning products in the water. You should be using environmentally safe cleaning products, body and hand soap, dishwasher soap, and laundry soap, anyway. You'll be able to use your greywater more safely. The idea is to reuse greywater for other uses that don't require drinkable/potable water.

The easiest use of greywater is to send it outside to water your garden. The next level is to treat the grey water enough so you can use it for non-potable uses, such as flushing toilets and maybe even faucets, dishwashers, and clothes washing. You can technically filter greywater and treat it to drink, but it might not make as much sense.

Reusing greywater reduces your water consumption from the water company and saves on your water bill.

CATCH RAIN

Freshwater rain falls on your house every year and flows into the sewer systems and rivers and back to the oceans. But you can capture that rain and use it. Then, you will need less water from the water company. To catch rain, set up rain gutters along your roof that funnel the water down into a rain catchment tank that sits next to your house. It is easy to set up the watering for your garden to use the rainwater in the tank. It's also easy to hook up a hose to wash your car, driveway, or even your dog. The next level is to set it up to use rainwater in your house for non-potable uses, like toilet flushing and clothes washing. Finally, you can use it for all water needs, such as sink water for washing dishes and even drinking and cooking water. Potable water must be filtered and treated properly before use.

You can calculate how much water you can collect per year by taking the average annual rainfall per year in inches in your area and the square footage of your roof. For every 1" of rain for 1 square foot of roof, you can collect .623 gallons. Multiply rain inches by square footage, and multiply by .623 to get the total gallons collected per year. Determining how often it rains throughout the year and your usage needs, you will be able to figure out how large of containers you need.

A bonus is that this is free water, so you will save on your water bill.

WALK OR BIKE

The most sustainable way to get around is to walk. It requires no products, so no energy used. Next is to bike. You just need to buy a bike, which takes resources but obviously not any energy to use it. Advocate for dedicated bike lanes. Advocate for walkable and bikeable communities. Get a home in a walkable/bikeable community. Live close to your work so you can walk or bike. Also, walking and biking is healthy!

USE PUBLIC TRANSPORTATION

If you can't walk or bike to where you want to go, use public transportation if possible. This means using the subway, trams, trains, or buses. Using public transportation uses fewer resources because you are sharing the ride with other people. Resources are used to build the train infrastructure and the trains, but then they are used by many people, instead of having to build individual cars for every person. If you use a car or take a taxi, you are using gas or electricity to move just yourself from A to B. When you share that ride with many other people, you use less gas or energy overall.

Try to live in an area where you can utilize public transportation. Encourage your government to build effective clean energy public transportation. We cannot continue to have a car-based culture. It consumes too much energy to build the cars and operate them. Having cars for us to share would be fine, but for our everyday activities, we should be walking, biking, or using public transportation and only using a car for trips that require it.

We also need to invest in a high-speed rail system to travel longer distances between cities instead of flying. Flying has one of the biggest impacts because it uses so much fuel.

DRIVE AN ELECTRIC CAR

If you can't rely on public transportation, walking, or biking and must have a car, then get an electric car. If you own a home, you can get a charging station setup at your home. If you have solar panels, you can charge it from that or opt in for clean energy from your utility provider for 100% renewable energy. If you don't have your electricity coming from clean renewable sources, then buying an electric car doesn't help much except to prepare you for having renewable energy in the near future, which can still help with the transition. If you rent, talk to your landlord about getting charging stations. There are some programs to set this up, and it should not cost landlords much, if at all. Or, offer to help cover the costs if that's a major issue.

Using an electric car allows you to drive without emitting CO_2 into the atmosphere, not only causing global warming but also pollution and VOCs in your region. Gas cars are the number-one air polluter in residential areas.

Most full electric cars have a 200–400-mile range. If you need more than that, then you can plan your trip around charging stations. If you think that is too inconvenient, you can buy a "plug-in hybrid" car that has a short all-electric range, roughly 30 miles, and uses gas for longer trips. This allows you to use mostly all electric for day-to-day trips, and when you drive on longer trips, you still have the convenience of gas.

If you can't get an electric car charger at your house for some reason, then at least get a hybrid car or any gas car with high miles per gallon. The less gas you use, the better. Also, buy a used car in this case. We can't be producing more all-gas cars and supporting that by buying new ones.

Note that replacing your gas car for an electric car doesn't mean that your gas car should be thrown away. Unfortunately, there is already "embodied carbon" in your car, which means energy and CO_2 emissions have been used to make that car. So, it does not make sense to just throw it away and replace it with a new car, since that will actually cause more impact on the planet. However, if you sell your car used to someone else, then that someone will use it until it reaches the end of its lifecycle. Still, buying a new or used electric car will make the demand for electric cars go up and help us achieve better infrastructure overall.

TAKE LONGER TRIPS INSTEAD OF MORE TRIPS

Flying in a plane takes a large amount of CO_2 emissions because of how powerful that jet engine needs to be to fly the plane. If you have a choice to drive or fly, driving can have less CO_2 emissions, especially if it is electric or has good gas mileage or you carpool. Trains and buses are even better. But the point here is that each trip you make takes energy for the transportation to get there and back. So, if you can take fewer trips overall, that would help conserve that energy. For vacations, if you go on one big trip for two weeks instead of three trips adding up to two weeks, then you would only consume one flight for the single trip. As far as business trips, just try to only travel to meetings if really needed. Take meetings with video conferencing tools so no travel is needed.

WORK FROM HOME

When you work from home, you don't need to travel to an office, so you save energy from that transportation. The more you can work from home, the less transportation energy you spend. You can use video and web conferencing tools to easily have group calls to maintain the benefits of collaboration. If your company requires in-person culture, then work from home one to four days a week.

WORK FOUR DAYS A WEEK

If you can do your job in four days for 32 hours a week instead of five days a week at 40 hours, then consider working four days a week instead of five. This way, you can save on the transportation energy and building energy that is used for a whole day per week, which would add up quite a bit if everyone adopted this. Other benefits would include a better work-life balance and happier and more productive employees. If you are an employee, make this case to your manager, and if you are managing a company, consider this new policy. There are studies that show that we are only productive part of the time when at work. But our culture has made it normal to work five days a week for eight hours per day. Are you really that productive every day for all hours? Most are not, depending on what exactly they do. But if you are a bus driver, for example, you drive a bus for your whole shift. This applies more to people working in an office.

With your extra day off, go for a hike and enjoy nature. Plant a tree. Do an activity that doesn't spend a large amount of resources or energy so it has a positive impact on the environment.

SUPPORT SUSTAINABLE GOVERNMENT POLICIES

Advocate, support, vote, protest, and promote government policies that support sustainable living goals. Do the same for political candidates who support those policies and will help implement them. Unfortunately, individual actions to make your own life sustainable will not be enough. There are many people who are not aware or do not care about living sustainably. The only way we can truly make enough change is through government regulation. We need regulation to ban the use of fossil fuels, single-use plastic, unsustainable farming practices, and more. The government needs to force the changes we need to make in order to protect our own existence. The government and our tax dollars need to be spent on restoration, regeneration, and sequestration projects in order to bring carbon out of the atmosphere and restore our environment to a healthy state.

Here are the top policies that government should focus on:

1. Convert to clean renewable energy.
2. Require regenerative farming practices.
3. Sequester carbon projects.
4. Design communities to be walkable and bikeable.
5. Provide easy public transportation on clean energy.
6. Make all new housing all-electric and energy efficient.
7. Regulate packaging to reduce single use and encourage reusable and compostable.
8. Require composting and recycling.

INVEST YOUR MONEY IN SUSTAINABLE BUSINESSES

Something that can get overlooked when trying to adopt a sustainable lifestyle is how you invest your money. If you have any savings, a 401K, a retirement fund, mutual funds, CDs, or stocks, then you have money invested in other businesses. If you are going to invest, put your money into businesses that support a sustainable world. Doing so supports these companies and helps them succeed. Unfortunately, many profitable companies earn their profits by extracting the earth's resources, damaging the environment, or utilizing inhumane labor. Where we invest our money drives the direction of our world. We should not invest our money into mining fossil fuels or cheap meat production, and instead, we should invest in solar energy, electric public transportation, regenerative farming, etc.

Invest in companies that create value to society; for example, a company that produces clothing or food. Other types of companies earn money on interest, like bank loans, or rent on land, like apartment buildings. These types of companies don't actually add value to society. You'll want to invest in companies that have a triple bottom line: profit, environment, and social. These companies seek to not harm the environment and provide socially just work places in addition to making a profit. Certified B Corps have these values.

Find opportunities to do impact investing. This is investing your money into a company that is beneficial for the environment or has a social impact while also making a profit. A safe Impact Investment is to put money into RSS Social Finance, which provides a guaranteed 1–1.2% return. They invest the money into organizations that make a good impact. CNote is another place to invest into good companies with slightly higher returns. You can also put your money into a local non-profit credit union, into a CD with a guaranteed return. The local non-profit credit union will keep the money in the local economy, which is better than your money going to big national banks and into dirty industries. Also, look for any other investment opportunities for organizations that are mission driven, with a purpose to help the world live in a more sustainable way.

WORK AT A JOB THAT SUPPORTS SUSTAINABLE LIVING

In addition to making efforts to live sustainably in your personal life, you can work at a company that also helps the world live sustainably. Work for a company that has practices to operate in sustainable ways.

A few example jobs:

1. Clothing company like Patagonia that is trying to adopt sustainable practices to produce clothes.
2. Farm that uses sustainable regenerative farming practices.
3. Any organization that helps protect the environment.
4. Construction company that builds using sustainable materials and practices.
5. Producer of electric cars, public transportation, or induction stoves.
6. Clean renewable energy developer, such as solar panels or wind turbines.

7. Restaurant or grocery store that sources sustainable food.

Even if a company is not doing something specifically that helps the world live sustainably, at least evaluate what the company does and how it operates to judge if you think they are having a bad impact on the environment, and choose not to be a part of that type of company.

BANK AT A LOCAL CREDIT UNION

When you use a bank account, you typically have money sitting in your checking and savings account. This is the money that is not invested. The bank will use the balances in people's accounts to invest for their own profit. You want to support a bank that invests in sustainable practices and provides value to society. Most big corporate banks invest money in real estate and extractive businesses like oil, logging, and cheap meat and farming. Open an account and move your banking to a local non-profit credit union. These are not for profit, so their goal is to serve the needs of its members. They also keep the money invested locally, which is better for your local community and economy. Most credit unions have the goal to consider the environment. So, it's better to use a credit union as your bank.

DONATE TO SUSTAINABLE ORGANIZATIONS

You can donate to many great non-profit organizations that are fighting hard to protect the environment or provide research and solutions to climate change and sustainable living. Here are a few of my favorites:

Sierra Club

This US non-profit is well known, well established, and trusted. They work to protect the environment and fight back against the government and corporations in the courts and lobbying activities. They try to prevent more oil drilling by stopping government leases, close down coal plants, protect national parks, and encourage governments to pass policies that protect the environment and implement environmentally friendly solutions.

Project Drawdown

This non-profit researches solutions to prevent climate change. They provide a breakdown of solutions and ranking for each one in terms of how much carbon it will remove from the air. Their focus is on removing carbon from the atmosphere to prevent climate change. Some of the solutions are changes for ongoing operation, and others are big projects that we can do to just take carbon out.

World Wildlife Foundation

WWF is a global non-profit organization that works to protect our nature. They work in 100 countries to protect oceans, forests, water access, food production, climate change, and wildlife species.

Greenpeace

This is a global non-profit fighting to protect our environment. They identify problems and crimes, and protest and communicate about them, as well as any solutions. They are more like activists and focus on communicating and promoting, not litigating like the Sierra Club. They are supported by donors and get no money from the government or corporations. They don't endorse political candidates, so they can stay focused on the mission.

350.org

Their focus is stopping the burning of fossil fuels to prevent climate change. They support clean energy solutions, help get bans on fossil fuels, and help remove financing to the fossil fuel industry.

National Forest Foundation

They plant trees for your donation. So much of our forests has been cut down and continues to be cut down. It is important that we plant trees to replace our forests to store the carbon out of the atmosphere.

HAVE FEWER KIDS

This is a complicated issue. I can't commit to what the right path is here. One thing I can say is that each human takes resources on the planet to just exist. The fewer humans we have, the less impact we have on the earth. Our population growth is getting to unsustainable levels. If everyone has one to two kids max, then at least our population would stop growing.

Now I'll provide a bit more information on this subject for consideration. There probably is a capacity to the Earth for the number of humans. However, that number adjusts if we want to give any space to any other living creatures or if we want to have it all for ourselves. The capacity also will change if we figure out more innovative ways to live more sustainably with the Earth, such as living in denser cities and producing food that takes fewer resources. We have been able to grow our population so dramatically due to using unsustainable, extractive processes, such as burning fossil fuels and using fertilizer from non-renewable sources and polluting our environment. It is fair to say that growing the population is unsustainable. We might already be past our capacity because we currently depend on extractive unsustainable practices to support our current population. Once we consume resources such as oil, fertilizer minerals, and wood, we might be in for mass adjustments. If we don't slow or reverse population growth, then we will have humanitarian issues down the road.

Keep in mind that if you are an eco-conscious human who is aware of these issues, it will be helpful for the next generation to have this awareness. Therefore, until the world is closer to living sustainably, it would be good to raise one to two children with these ideals to hopefully help humanity meet these goals. This is an alternative to having no children because you are concerned about the impact.

I feel that it is unnatural to restrict reproduction. Everything living in nature seems to attempt to reproduce to the limit typically imposed by other forces. So, we could continue in this fashion, as we have, to allow all to reproduce as much as they want. But then, we would have to allow natural forces to stop populations from reproducing, which could include lack of water or food. But if we don't want to see suffering, then we should not allow populations to get to this point.

Let's go back to the question of capacity. Even then, there are different levels. If you have a limited space, such as 100 acres of land but only five people live there, then each person has a much more enjoyable life with more resources. Maybe that space can support 1,000 people, but everyone will be very poor because they need to share limited resources. Then it becomes a matter of who gets more or better resources than another.

For example, let's take a look at a city that has grown in population because of unsustainable practices using oil and fertilizers to grow food and use up groundwater. They grew a population, ran out of oil and groundwater, and must live within their regional resources. Everyone is poor and life is hard, if people are able to live at all. Some of these people want to move to a city in a country that can still support more population, but maybe that area has a comfortable population where they have a good quality of life. To increase the population would downgrade the quality of life. What do you do about that?

I hope that we have the will to manage this before it causes suffering and difficult choices.

SUSTAINABLE CONCEPTS
YOU SHOULD KNOW

PARIS CLIMATE AGREEMENT

The Paris Agreement is a climate change agreement by the United Nations, signed in 2015 by 195 countries. The agreement is to keep the average global temperature well under 2° Celsius above pre-industrial temperatures and to attempt to limit the increase to 1.5° Celsius. Each country is to define their own plan on how to reduce global warming within their country and report on their progress. Developed countries are supposed to provide financing to undeveloped countries to help them with climate change solutions. This agreement is great and provides a general understanding of the goal and that all countries must be involved, but many think that it is not strict enough on implementation and enforcement. It is a good step, but given the urgency reported in the IPCC report, we will need much more action.

IPCC CLIMATE 1.5 DEGREE REPORT

In the Paris Climate Agreement adopted in 2015, they asked the IPCC (Intergovernmental Panel on Climate Change, a part of the United Nations) to create a report on the impact of global warming if our planet's average temperature was 1.5° Celsius above pre-industrial temperatures and warmer. Ninety-one authors and editors were selected from forty countries to draft the report. These authors were nominated globally for their expertise in the topic. The panel researched this topic, peers reviewed their drafts, and they published their findings in 2018. The report claims that we need to not go above 1.5° Celsius; otherwise, it will be dangerous and cause myriad problems. Sea levels will rise, which will force people to move inland or build new infrastructure. More extreme weather like heat waves, hurricanes, droughts, floods, and fires, will be more common. Fast changes and swings in temperatures can also cause species to be unable to adapt and go extinct, making our ecosystem less resilient.

We reached 1° Celsius above pre-industrial levels (1900 AD) in 2017 and are warming at an average rate of 0.2° every 10 years. So, it's estimated that we will reach 1.5 by 2030. We need to cut our current GHG (greenhouse gas) emissions by 45% by 2030 and become net zero by 2050 in order to remain no more than 1.5° Celsius above. GHGs comes from burning fossil fuels like gas in cars, factories, producing products, building homes, running A/C and heating, etc. In order to cut our emissions, we need to reduce our use of energy and convert to non-fossil-fuel energy sources. In addition, we need to implement carbon dioxide removal projects that will remove and store CO_2 from the atmosphere, such as planting trees.

The important thing about this report is the wide consensus of scientists who all agree on the data, how urgent we need to make major changes to our current systems to reduce our emissions, and the major consequences if we don't. We have the knowledge and solutions now to solve this; all we need is the political will.

This report is what is driving policy in government and has brought awareness to the public to demand change.

SUSTAINABLE DEVELOPMENT GOALS (SDGs)

There are seventeen Sustainable Development Goals adopted by all United Nations member states in 2015 in order to peacefully prosper for the benefit of all humans. These goals recognize the need to protect our environment, prevent climate change, end poverty, and reduce inequality. The UN Division for Sustainable Goals works to implement these goals by 2030 with all countries. Here are the goals:

1. End Poverty
2. Zero Hunger
3. Good Health and Well-Being
4. Quality Education
5. Gender Equality
6. Clean Water and Sanitation
7. Affordable and Clean Energy
8. Decent Work and Economic Growth
9. Industry, Innovation, and Infrastructure
10. Reduced Inequalities
11. Sustainable Cities and Communities
12. Responsible Consumption and Production
13. Climate Action
14. Life Below Water
15. Life on Land
16. Peace, Justice, and Strong Institutions
17. Partnerships for the Goals

PROJECT DRAWDOWN

Project Drawdown is a non-profit organization that has done extensive research to identify the top solutions to prevent climate change. Basically, the goal is to reduce carbon emissions, as well as remove carbon and store it from the atmosphere. The org works with governments, businesses, universities, and philanthropists to deploy climate solutions. It ranks its top eighty solutions in order of how much CO2 they will reduce. There is a book you can read or view on their website.

DOUGHNUT ECONOMICS

Doughnut Economics is a term that economist Kate Raworth coined on how to look at our economy. She wrote a book explaining the concept. Our current economy looks at Gross Domestic Product (GDP), imports, exports, and stock prices. We only look at sales and producing more and more. The more we sell and produce, the better. Our current view does not consider the fact that we are extracting our natural resources and damaging our planet. We will deplete our resources in the future if we continue on this track.

Kate provides a way to look at our economy within the planet's boundaries. The "doughnut" is from the shape of a diagram that she designed, which shows a circle with the outer boundaries of our planet's resources. If we don't learn to live within the earth's ecological carrying capacity, then we will cause climate change, pollution, and biodiversity loss. The doughnut itself is where we want to live, where we have clean water, food, energy, housing, etc. in balance with nature. It is important that we change the way we view our economy because this drives our culture and activities. We need to recognize our ecological limits in order to have a sustaining society long term.

CIRCULAR ECONOMY

Right now, our economy is a linear extractive economy. We extract resources from the environment, produce products, and sell them to consumers. Consumers then throw them away and we put them in a landfill. You can observe in nature that everything is part of a cycle that creates balance. For example, a tree grows and uses the healthy soil. When it dies, it gets decomposed by mushrooms and insects and microorganisms that break the organic matter back down into the healthy soil for more trees to use and grow. When you cut down a tree and take it out of the forest, the nutrients, organic matter, and carbon in the tree don't return to the soil, and therefore, the soil doesn't retain the nutrients.

Another example is that rain falls from the sky, flows into rivers, then back into the oceans. Then, warm weather evaporates the water from the oceans back into clouds, and they move toward mountains and rain again. The water is never really gone—it just moves from one form to another and from one place to another in a balanced cycle.

We need to think about our economy in the same way. We need to take resources such as trees for wood, use them to build furniture and housing, and then when the wood is old and needs to be replaced, we need to return the old wood back to the forests to put nutrients back into the soil to grow more trees. The biggest change here is thinking about how to move the resources back to the beginning to create a circular system.

Another example is that we grow food on a farm using healthy soil and create produce to eat. We eat produce, but we can then put the scraps from that food into the compost bin. When we compost organic matter, like food scraps, microorganisms break that food down, just like the trees that fall in the forest, and we use it as natural fertilizer on the farm to grow more food. We return the resources back to the beginning.

To take it further, food is eaten by humans, and then we digest it and poop and pee out whatever we did not retain. We need to take that human poop and urine and recycle it back into our environment to use for growing more food. For everything we produce, we need to think about the full life cycle of the product: what its source materials are, what its useful life is, and what happens after it's no longer useful. We should not be putting anything into a landfill.

We need to have this circular approach in the design of products and systems. If we don't figure out how to do this, we will eventually be forced into a situation where we deteriorate our resources and make it more difficult to restore a healthy balanced ecosystem.

GREEN NEW DEAL

The Green New Deal is a government plan to invest heavily into converting our public infrastructure to be more environmentally sustainable. This mostly focuses on electrifying the nation; moving our energy to clean, renewable, non-fossil fuel sources; making buildings more energy efficient; and moving to electric vehicles and high-speed rail for transportation.

The goal is to prevent climate change, as warned by the IPCC report, and to implement all the action, policies, investment, incentives, and regulation that would be required to achieve that goal. The name comes from Franklin D. Roosevelt's "New Deal," which was a series of reforms and public works projects in the 1930s to improve the US infrastructure and guide the US out of the Great Depression. It was a considerable investment by the federal government to help the country make a big transition. Now, with the awareness of climate change and the considerable problems it can cause if we don't act within the next 12 years, we need this kind of major investment and transition. Right now, the Green New Deal is a concept, with a draft and different presidential candidates in the 2020 election with their own versions. Some sort of Green New Deal needs to still be passed in Congress.

CLIMATE EMERGENCY

The IPCC report in 2018 states that we must reduce our CO_2 emissions by 45% by 2030 and be net zero by 2050 in order to not go above 1.5° Celsius above pre-industrial levels. Otherwise, we will have major consequences. Given the major changes needed within 12 years—which is a very short amount of time—some say that we need to declare a climate emergency and act accordingly. Even though we are already seeing the effects of climate change with sea level rise, heat waves, and fires right now; climate change is not affecting the majority of people yet. Many feel it will not affect them in their lifetimes. But communities who are affected now or who understand and believe in the science reported can understand that there is an emergency, and we need to treat it as one.

Declaring a climate emergency makes the government put in more intensive immediate resources to solving climate change. It is a way to look at climate change as critical.

ECOLOGICAL FOOTPRINT

Global Footprint Network has developed a method to measure our ecological footprint. Everything we produce and consume and build takes resources from our planet, such as wood, fish, etc. Everything is calculated down to how much land is required. Every building requires land, every farm requires land, and producing energy requires land—wind and solar, for example. When activities put CO_2 in the atmosphere, we can remove and sequester it, such as by planting trees, but that also requires land. Everything can be calculated to require some acres of land. That is how you measure and report the ecological footprint. The goal is to standardize how we measure our footprint so we can track it globally for every city, country, business, etc. to enforce any regulations and track progress toward our goals.

We only have one planet for the global population, and currently, the world lives as if it has 1.75 planets. We will need to reduce our ecological footprint to live within one planet. Each year, Global Footprint Network records when we overshoot our one planet resources for the year. In 2018, for example, we overshot our resources by August. This is called the Earth Overshoot Day. That means we are using up existing stock, like cutting down trees, or putting more CO_2 into the atmosphere than our one planet can replenish or balance. As we continue this overuse of our planet, we will eventually deplete existing resources and face the consequences.

On their website, you can calculate your own ecological footprint and see other reports of the current ecological footprints of specific countries and cities.

There are a few other organizations that have developed a footprint calculator. Another common method to calculate your footprint is to measure your *carbon* footprint, which indicates how much CO_2 a product or your life emits. For example, eating beef makes more CO_2 emissions than plants. Measuring in terms of carbon is easy to understand and valuable, but I don't think it can be applied to all things. In the longer term, if we look at becoming net zero and not burning fossil fuels, this measurement will become less useful.

We need to make this measurement standard and transparent and report on this every day, every quarter, every year, just like we do our GDP and stock market value, to bring more awareness and track progress. I would like to see the ecological footprint number on every product I buy and every meal I order from a restaurant to help me understand and make sustainable decisions. I'd like to see my own ecological footprint, as well as those of my building, my city, and my country, and compare it with others.

LIFE CYCLE ANALYSIS

Every product produced is made of raw materials that are produced, transported and consumed, then disposed of and hopefully, recycled or composted. You can perform a life cycle analysis on a product to determine the environmental impact it has. This can also be called an

Environmental Product Declaration (EPD). You look at the product's full life cycle, from the raw resources extracted to create it to when it has been fully recycled or returned to the environment. Consider the amount of materials used, the energy used, possible reuse or recycle options, and any harm to the environment.

It is great to do this analysis when comparing different options for products, so you can choose the one that causes the least environmental impact. For example, after a study of life cycle analysis, some researchers found that it was actually better for the environment to distribute tuna fish in lightweight foil packaging instead of metal cans. Metal cans can be recycled, but they are heavier for transportation, which requires more energy. Add that to the energy it takes to recycle the cans, and you have an overall larger impact on the environment than using foil pouches, which are not recyclable but can be safely disposed in a landfill. This is why it is so important to do a full life cycle analysis so you truly know the best solution rather than assuming what seems better.

Typically, we think of the life cycle of a product as cradle-to-grave, meaning that it ends up in a landfill. But we need to change our thinking to cradle-to-cradle in order to create a product where nothing ends up in a landfill, and instead, all materials are returned to the environment to support the next product.

BUILDING CERTIFICATIONS

There are a few organizations that provide certifications for buildings that are built and operated more sustainably. Just like food getting certified as organic, humane, or non-GMO, it allows consumers to know that the building has met certain guidelines and requirements, influencing their decision to buy it. Developers use these certifications because they want to do good things for the environment but it also allows them to market the building to environmentally conscious consumers. An individual can even get a certification for their own home to know that they are living in a home that is more sustainable and to prove that for the next buyer. More recently, governments are adopting building development codes that require certifications or have similar requirements.

However, there is still a wide difference between basic building codes and how sustainable you can build and operate a building. Certifications for buildings mostly look at the materials used to build the building, if the appliances being used are electric, and how much energy it takes to operate the building. Reducing the energy used in materials and operating a building takes many different things into account. For example, you can orient your building to reduce sunlight heating the interior or add more insulation to require less heating.

Here are the current certifications to consider:

LEED

This certification is the longest standing and most well-known worldwide. They have different levels of certification, from silver to platinum. You can get certified for the construction of the building itself and continued operation. It provides certifications for buildings, as well as communities or cities. The certifications focus on reduced energy, water, and waste as well as human health improvements. They also have a Net Zero certification.

Living Building Challenge

This organization has a few different certifications for buildings, communities, and products. Its primary Living Challenge certification focuses on the areas of place, water, energy, health and happiness, materials, equity, and beauty. The org also has certifications Zero Energy and Zero Carbon.

Well

This certification is focused more on the quality of life of humans in buildings and communities, thinking about air, water, nourishment, light, movement, thermal comfort, sound, materials, mind, and community. The focus is more on the health of the building and comfort and less on the environment or sustainability.

FIRST NATION WISDOM

First Nations include Native Americans and any other groups of people who inhabited a place before they were conquered or colonized. These were the civilizations that lived in the area for thousands of years prior to the industrial revolution. First Nations survived in their land by learning it well and figuring out how to live in balance. Every territory has different weather, animals, plants, soil, systems, patterns, etc. Since Europeans colonized the Americas only a few hundred years ago, we don't have the long history and knowledge of this land. We also have fully adopted an industrialized culture and processes. With our efforts to live more in balance with the planet and live more sustainably, the First Nations can provide knowledge of how to do that since they have been doing that for thousands of years. We can learn from them and their way of life to help with these efforts. Some First Nation cultures lived in balance with the land, but there were also some that did not and perished because they depleted their resources. We can also learn from the civilizations before us that collapsed.

CARBON OFFSETS

When you fly a plane, drive a car, or use your air conditioner, you are responsible for emitting carbon into the atmosphere. This is due to burning fossil fuel for energy.

Carbon is also embodied in many products because of the energy used to produce it. The idea of carbon offsets mostly comes from the energy expended for fuel burned, like for an airline flight. When you buy a flight now, there could be an option to purchase carbon offsets with your ticket. They calculate the CO_2 you are responsible for emitting for your flight, then allow you to purchase offsets to balance it.

To buy carbon offsets means to invest into projects that will remove carbon (CO_2) from the atmosphere. The most common type of project is planting trees, since trees will grow and carbon is stored in the tree trunk. But there are other methods that pull carbon out of the atmosphere. For example, taking a flight from San Francisco to New York will emit 0.688t CO_2. You can then buy carbon offsets from an organization that would allow them to pull the same amount of carbon out of the atmosphere—in this case, roughly $20. You can apply the same concept and calculate your carbon footprint for your life each year and purchase carbon offsets for everything. Or you can just donate to these types of projects in general.

It is important for us to put a price on carbon to understand the real cost of everything we do because of global warming. We are burning too much fossil fuel, and the atmosphere is getting too much carbon in it and is causing the planet to warm. It is important that we understand and pay the actual costs for emitting CO_2, since it's an issue now. Buying carbon offsets is a voluntary thing, but it's nice that it is bringing some awareness. But the real solution is for there to be an actual fee or tax on carbon. The government will need to play more of a role in making sure we are removing as much carbon as we emit. What's better than buying a few dollars of carbon offsets is donating to these types of projects. Better still, advocate and vote for policies that will actually create a carbon tax. The best idea I've heard is to put a tax at the source of the carbon because it is easy to regulate.

However, be cautious of the idea that you can just buy carbon offsets and feel good about all the emissions you are responsible for. It certainly feels easy and can definitely help. But in the end, we do need to execute those carbon removal projects, and there are a few limitations there. For example, if you donate to have trees planted for your carbon offsets and everyone in the world did the same thing, there is limited land available to plant all those trees. Also, we would need to make sure those trees don't eventually get cut down. Every carbon removal project requires something. The most important thing is to reduce our carbon footprint instead of just thinking that we can throw money at the issue. There is still only one planet, and we need to live in a balance with it. We can't just keep increasing our carbon footprint, then spending money to offset it.

PERMACULTURE

Permaculture is a set of design strategies to think about the whole system mostly related to growing food. You try to design a garden, farm, or food forest to be regenerative, organic, and a circular system. The concept also relates in general to building a system in balance with nature, using natural systems, and sustaining a community.

Common Ideas in Permaculture

1. Compost food scraps and animal manure to use as fertilizer.
2. Catch rainwater and hold in a pond or holding tanks.
3. Use cover crops to return nutrients to the soil.
4. Plant different types of plants that use and provide different elements from the soil that then benefit all the plants together.
5. Plant different species of plants with different heights so they can be grown together in smaller space.
6. Develop a food forest. Plant and grow food-providing plants in our cities and gardens.

BIOPHILIA

Biophilia is the idea that humans are drawn to nature. When we design a building or anything similar, we should look to nature for inspiration and try to design something that makes it feel natural or connects people with nature. For example, you can have wide open sliding walls that open up to the outside and see the forest. You can create an indoor garden wall to have many plants inside a building. Or you can design a couch to look like a leaf. You could use more natural materials, such as wood and stone, instead of plastic. A not-biophilic design, for example, is an office building with fluorescent lights in a carpeted room, with no plants and all rectangle cubicles. Or it's a city with concrete and buildings and not many trees or parks.

The idea is that being in nature is a normal state for humans, and therefore, we feel better being in nature. So, we should try to keep our built environment feeling more connected with nature.

B-CORP

B-Corp (Benefit Corporation) is a type of corporation that you need to get your business certified. Most companies' only goal is to make a profit. This is not the case for non-profit organizations, but these still have a mission that drives them. A B-Corp must take into account the triple bottom line, which is People, Planet, and

Profit. It should do good for people by taking care of its employees and not exploiting them, while also taking care of its community. It takes into account the planet by looking at its environmental impact and trying to reduce that. But it still has the goal to make a profit, which is what makes it different from a non-profit, for example.

A company must get certified to advertise the seal that you can see on products, with a letter B. If you see that certification on a product, you know that that company is likely operating better for the planet than other companies, so you should favor B-Corporations.

REWILDING

Rewilding means to take an area that was previously built on and controlled by humans and get it back to a wild natural place. For example, let's say that you have an area that used to be an industrial area, with a bunch of warehouses that are old and unusable. Instead of tearing those buildings down and building new ones, you decide to remove the buildings and restore the nature by planting native species. You might need to do a bit of terraforming to get that land back to what it was. The goal is to let nature take over the land and manage itself. Unlike a park that is planted and organized and controlled by trimming plants, watering them, and creating paths, to rewild an area is to *not* manage it and just let it be wild.

Humans have taken over a large amount of land for our own use and have left only a bit of it wild with some national parks. The goal is to build our communities denser and leave more surrounding areas wild so other species can also exist. Certain cities, like Portland, are modifying their regional plans to rewild areas of the county and remove human control and buildings instead of building out and over more land.

UPCYCLE

Upcycling is to reuse materials that would have been thrown away. This could be by-products from an industry, such as scraps of cloth from a clothing factory, saw dust from a wood cutting mill, containers used for shipping that are not needed after the shipment was received, products that are worn out and in bad shape, or any unwanted products by anyone. Then someone gets these waste products and transforms them into new products.

Upcycling is not the same as recycling or down-cycling. To downcycle is to recycle and convert materials into lesser value materials. For example, plastic bottles can be melted down and made into new plastic that might be of lesser quality than virgin plastic so you might use them mixed with other materials in things like carpet. In downcycling, you smash and grind or melt and try to use the base material but also spend energy doing it, and it typically is of lesser quality than the original product.

Upcycling, on the other hand, is like taking the plastic bottles and creating a piece of artwork out of them without altering them much. Using reclaimed wood is upcycling: you take old wood from another building and use it in a new building with minor improvements. To downcycle or recycle that wood would be to grind it down into a sawdust and create particle board, which is not as strong and takes a great deal of energy to make.

Another example is secondhand clothing, which typically just needs to get washed and lightly repaired to be in good enough condition to be sold again. Compare this to the energy required to break down that fabric and recycle it to spin new fabric completely.

YOUR GOOD HUMAN SCORE

We have developed a series of questions so you can calculate your "Good Human Score," which indicates your level of living sustainably with your available options. Depending on where you live, some things you can't adopt. However, in the long run you have a choice and could adopt every action to live the most sustainable lifestyle.

Each question has options to select from, and each option has a point value. The more points you earn, the better. Questions are weighted, so if something provides more value to living sustainably, it will be worth more points. The maximum amount of points you can score is 200, which means you are doing everything indicated to live in the most sustainable way.

Go to the website, goodhuman.eco, for the most current calculator and to fill out these questions in a more user-friendly form. You can also save your results and continue to update as you make progress toward living more sustainably.

ANSWER THE QUESTIONS BELOW

Read each question and write down the points earned in the space provided. Then add up your points to get your score at the bottom. Use a pencil so you can update your answers over time. The questions are roughly in order of the topics in this book, so you can reference it for further information. If something does not apply to your situation, give yourself the points, anyway. For example, if you don't have AC, you can take the points for having an energy-efficient AC.

1. Do you get your energy from clean renewable sources?

If you have opted in for 100% renewable energy from your utility company or if you have solar panels or wind turbines on your property that provide 100% of your energy, then answer yes. Otherwise, say no.

Put 20 points if yes, 0 points if no.

Points Earned: _____

2. Is your home all electric?

Are all your appliances powered by electricity instead of fossil fuels?

1. Water Heater
2. Heater
3. AC
4. Stove
5. Oven

1 point for each appliance, 5 total possible.

Points Earned: _____

3. Are your appliances energy efficient?

They should be Energy-Star certified.

1. LED Lights
2. Electric/Induction Stove
3. Oven
4. Refrigerator
5. A/C
6. Heater
7. Water Heater
8. Washer/Dryer
9. Dishwasher
10. TV

1 point for each appliance, 10 points max.

Points Earned: _____

4. Is your house efficient for heating and cooling?

This means that it does not have any air leaks, it is well insulated, the windows are double paned, and sun and shading is used to manage the desired temperature without using more AC/heating. If you are not sure, you can do an evaluation.

Give yourself 0 to 3 points depending on your rating.

Points Earned: _____

5. Do you use the least amount of energy that you can?

Do you avoid wasting energy by not leaving things on, such as lights or appliances? Do your devices auto shut off when not in use? Do you set your AC/heater temperature ranges the lowest possible but still comfortable enough?

Give yourself a score from 0 to 3.

Points Earned: _____

6. Is your home made of sustainable materials?

Depending on the percentage of home materials, give yourself a score between 0 and 10.

Points Earned: _____

7. Do you eat a small amount of meat?

10 points if you eat less than 8 oz. every 4 weeks. 0 points if you eat meat every other day. Or choose a score in between depending on where you think you fit.

Points Earned: _____

8. Do you waste food?

10 points if you actively try not to waste any food: you are careful to buy what you need and consume it before it spoils, and take restaurant leftovers to eat at home. 0 points if you often throw away food. Or choose a score in between depending on where you think you fit.

Points Earned: _____

9. Do you eat sustainably grown food at home?

20 points if you go out of your way to buy sustainable regenerative grown food whenever possible, 10 points if you buy organic food. 0 points if you buy conventionally grown food. Or choose any score in between depending on where you think you fit.

Points Earned: _____

10. Do you eat at restaurants that source their food from sustainable sources?

2 points if you do this most of the time and whenever possible. 0 points if you don't try. Or choose any score in between depending on where you think you fit.

Points Earned: _____

11. Do you buy only products you need?

Buy good quality that will last, with purpose, with the goal to buy the fewest products possible.

Give yourself 0 to 5 points depending on what you think.

Points Earned: _____

12. Do you repair products when possible instead of re-placing?

1 point if yes, 0 if no.

Points Earned: _____

13. Do you buy used products when possible instead of new?

1 point if yes, 0 if no.

Points Earned: _____

14. Do you sell or gift your stuff instead of throwing it in the trash if possible?

1 point if yes, 0 if no.

Points Earned: _____

15. Do you buy sustainably made clothing?

2 points if you make a big effort to buy if possible. 1 point if sometimes. 0 if no consideration.

Points Earned: _____

16. Do you buy sustainably made products?

Also consider if the packaging materials are sustainable.

2 points if you make a big effort to buy if possible. 1 point if sometimes. 0 if no consideration.

Points Earned: _____

17. Do you use reusable shopping bags?

1 point if yes, 0 if no.

Points Earned: _____

18. Do you use reusable containers?

1. Water bottle

2. Coffee cup
3. Lunch tupperware/wrap, picnic plates, forks, knives, spoons
4. Refillable dish soap, bath soap, hand soap, and laundry soap
5. Refillable food containers: milk, juice bottles, nuts, grains, etc.

1 point each, 5 points max.

Points Earned: _____

19. Do you use rechargeable batteries instead of single use?

1 point if yes, 0 if no.

Points Earned: _____

20. Do you use reusable napkins/towels instead of paper towels?

1 point if yes, 0 if no.

Points Earned: _____

21. Do you buy and receive sustainable gifts?

1 point if yes, 0 if no.

Points Earned: _____

22. Do you use eco-friendly cleaning products?

1. Bath soap
2. Shampoo
3. Conditioner
4. Hand soap
5. Dishwashing soap
6. Dishwashing detergent
7. Laundry soap
8. Glass cleaner
9. Counter cleaner
10. Floor cleaner

1 point each, 10 points max.

Points Earned: _____

23. If you have clothing that contains polyester or other synthetic plastic materials, do you wash them with a micro plastic filter installed or inside a bag to prevent the

plastic from leaking into our oceans?

1 point if yes, 0 if no.

Points Earned: _____

24. Do you compost your food scraps?

1 point if yes, 0 if no.

Points Earned: _____

25. Do you recycle properly?

1 point if yes, 0 if no.

Points Earned: _____

26. Is your human poop and pee utilized for anything by your sanitation facility?

1 point if yes, 0 if no.

Points Earned: _____

27. Do you have water-efficient appliances?

Do you also attempt to conserve water usage?

1. Toilet
2. Dishwasher
3. Faucets
4. Shower head

1 point for each, 4 points max.

Points Earned: _____

28. If you have a garden or landscaping, does it require a lot of watering?

Do you have more native plants that don't require watering? If you must water, do you water efficiently using drip irrigation?

2 points if your garden doesn't require much watering. 1 point if your garden requires watering, but you water it efficiently. 0 points if you need to water a lot, don't water efficiently, or have a lawn.

Points Earned: _____

29. Do you utilize your grey water?

1 point if yes, 0 if no.

Points Earned: _____

30. Do you catch rain?

1 point if yes, 0 if no.

Points Earned: _____

31. How do you get to work?

10 points if you work from home or walk or bike to work.
9 points if you use public transportation that is electric.
6 points if you use public transportation that is not electric.
5 points if you drive an electric car that is charged by clean energy.
2 points if you drive a car not charged by clean energy up to 20 minutes one way.
0 points if you drive a car for more than 20 minutes one way.

Points Earned: _____

32. What car do you own?

10 points if you don't own a car.
7 points for an electric car charged by clean energy.
5 points for an electric car not charged by clean energy.
3 points for a hybrid car.
2 points for a gas-efficient car of 30MPG.
0 points for a gas car.

Points Earned: _____

33. How much do you fly?

1 point if you fly only when absolutely needed and try to limit your trips in favor of longer trips for vacations each year. 0 points if you don't consider flying less.

Points Earned: _____

34. Do you support sustainable government policies and candidates?

30 points if yes, 0 if no.

Points Earned: _____

35. Is your investments in sustainable businesses?

If you don't have any investments, answer how you would invest if you could.

5 points if 100% invested in sustainable business, 0 points if 0% or any points in between depending on percentage.

Points Earned: _____

36. Do you work in a job that supports sustainable living?

5 points if yes, 0 if no.

Points Earned: _____

37. Do you bank at a local credit union instead of a big national bank?

1 point if yes, 0 if no.

Points Earned: _____

38. Do you donate to organizations that help the world

live more sustainably or protect the environment?

1 point if yes, 0 if no.

Points Earned: _____

39. Do you have or plan to have 2 or fewer children?

1 point if yes, 0 if no.

Points Earned: _____

Now add up all the points you earned above to get

YOUR GOOD HUMAN SCORE

_____